OUR SOLAR SYSTEM

EARTH

by Alissa Thielges

AMICUS

water

mountain

Look for these words and pictures as you read.

satellite

moon

Earth is our home.
It is a planet.

Earth goes around the sun.
It takes about 365 days.
It is the third planet from the sun.

See the water?
We need it to live.
No other planet has this.

water

See the mountain?
It is the tallest on Earth.
People climb it.

mountain

See the satellite?
It goes around Earth.
It tracks the planet's changes.

satellite

See the moon?
It is close to Earth.
Astronauts have walked on it.

moon

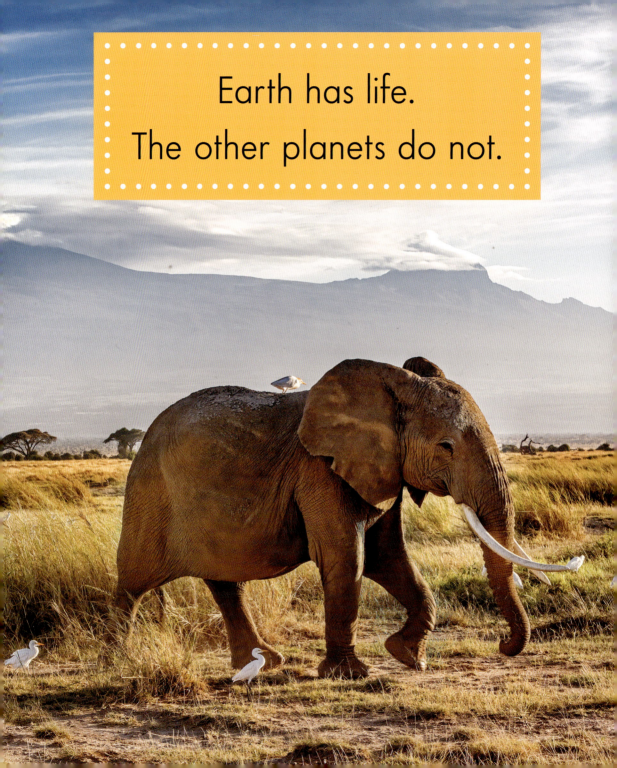

Earth has life.
The other planets do not.

water mountain

Did you find?

satellite moon

Spot is published by Amicus Learning, an imprint of Amicus
P.O. Box 227, Mankato, MN 56002
www.amicuspublishing.us

Copyright © 2024 Amicus. International copyright reserved in all countries. No part of this book may be reproduced in any form without written permission from the publisher.

Library of Congress Cataloging-in-Publication Data
Names: Thielges, Alissa, 1995– author.
Title: Earth / by Alissa Thielges.
Other titles: Spot. Our Solar System.
Description: Mankato, MN : Amicus, [2024] | Series: Spot. Our Solar System | Audience: Ages 4–7 | Audience: Grades K–1 | Summary: "Simple text and a search-and-find feature reinforce new science vocabulary about Earth's geography and space features for early readers"—Provided by publisher.
Identifiers: LCCN 2022035871 (print) | LCCN 2022035872 (ebook) | ISBN 9781645492702 (library binding) | ISBN 9781681527949 (paperback) | ISBN 9781645493587 (ebook)
Subjects: LCSH: Earth (Planet)—Juvenile literature.
Classification: LCC QB631.4 .T454 2024 (print) | LCC QB631.4 (ebook) | DDC 525--dc23/eng20230106
LC record available at https://lccn.loc.gov/2022035871
LC ebook record available at
   https://lccn.loc.gov/2022035872

Printed in China

Rebecca Glaser, editor
Deb Miner, series designer
Lori Bye, book designer
Omay Ayres, photo researcher

Photos by Flickr/NASA, cover; Getty/ewg3D, 4–5; Shutterstock/Digital Photo 6–7, Jane Rix 14, Johan Swanepoel 10–11, Klagyivik Viktor 12–13, Vixit 8–9 Vladi333, 3; Wikimedia Commons/Joshua Stevens NASA Earth Observatory, 1, 16

EARTH